Little Lessons For Kids
presents

Chili's Multiplication Party

1 x 3 = 3

2 x 3 = 6

3 x 3 = 9

4 x 3 = 12

Improve
your
multiplication
skills

Chili the
Little Big Dog

Little Lessons [illegible]
Visit us at www.little[illegible].com
Published by Little Lessons [illegible] Publishers.

-Multiplying by 1 is a breeze! The answer is always the number being multiplied by 1!

Examples: 1x2=2 1x3=3 1x4=4 1x5=5

-When you multiply by 2, the answer is always an even number!

Exmples: 2x1=2 2x2=4 2x3=6 2x4=8 2x5=10 2x6=12 2x7=14 2x8=16 2x9=18

-Multiplying by 5 is easy!
Each answer ends with 5 or 0
Examples:
5x1=5, 5x2=10, 5x3=15,
5x4=20, 5x5=25, 5x6=30
5x7=35, 5x4=40, 5x9=45

INSTRUCTIONS

Copy the number combinations into the puzzle by matching numbers. Then connect the dots.

Example: multiples of 3

1 • 3
2 • 6
3 • 9

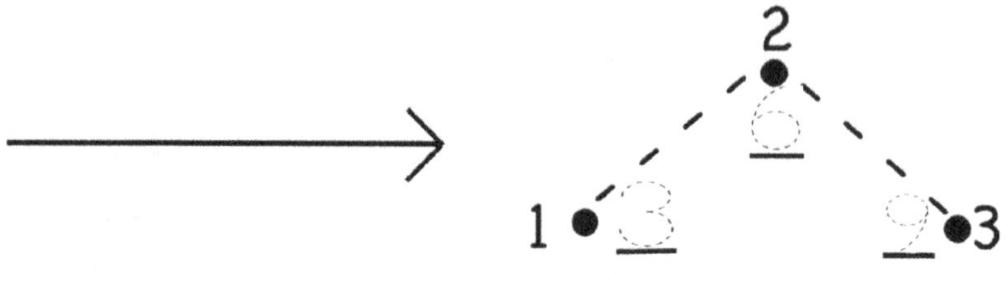

Each line starts at 1 and ends at 9.

Once you finish a line, connect the dots before moving on.

Number Combinations

x3

1 • 3
2 • 6
3 • 9
4 • 12
5 • 15
6 • 18
7 • 21
8 • 24
9 • 27

Example

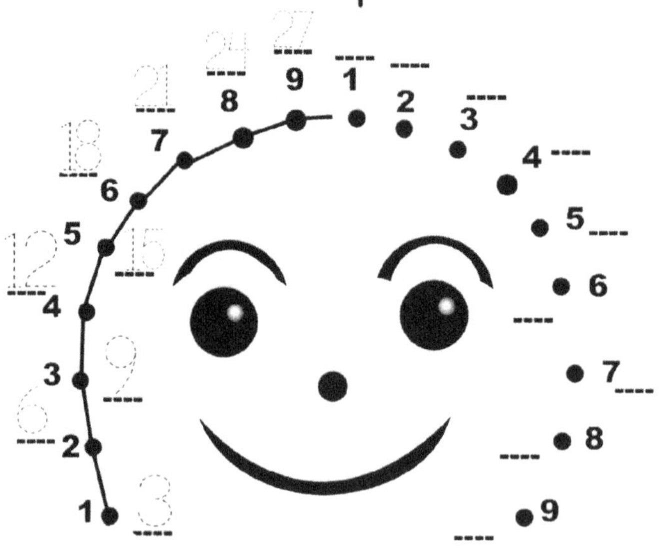

Video examples available at www.littlelessonsforkids.com

x3
1 • 3
2 • 6
3 • 9
4 • 12
5 • 15
6 • 18
7 • 21
8 • 24
9 • 27
1. SEE THE NUMBER COMBINATION
2. TOUCH THE NUMBER COMBINATION
3. SAY THE NUMBER COMBINATION
4. WRITE THE NUMBER COMBINATION IN THE PUZZLE

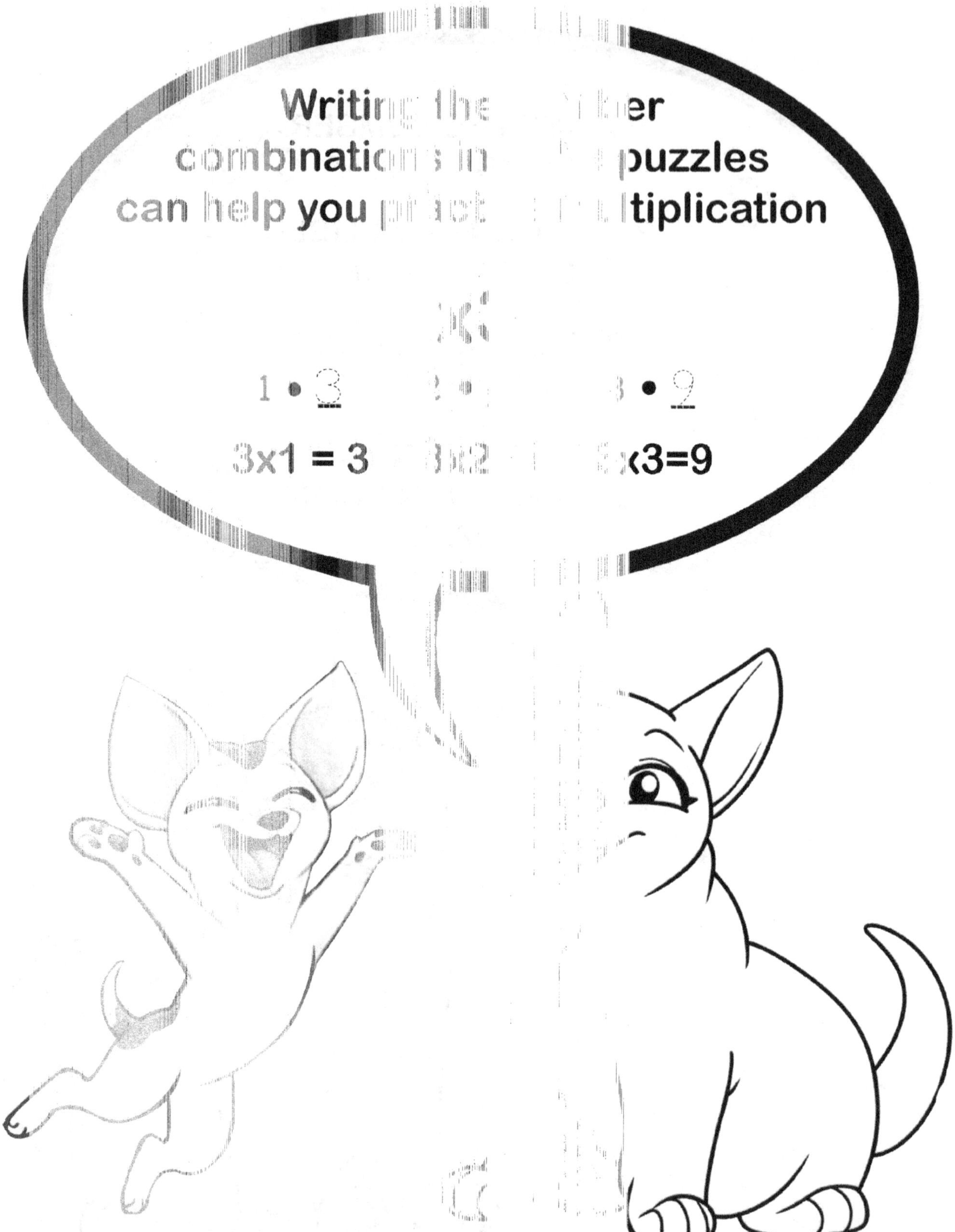
Writi the er
combinations in puzzles
can help you tiplication
1 • 3
3x1 = 3
• 2
x3=9

3

WARM UP

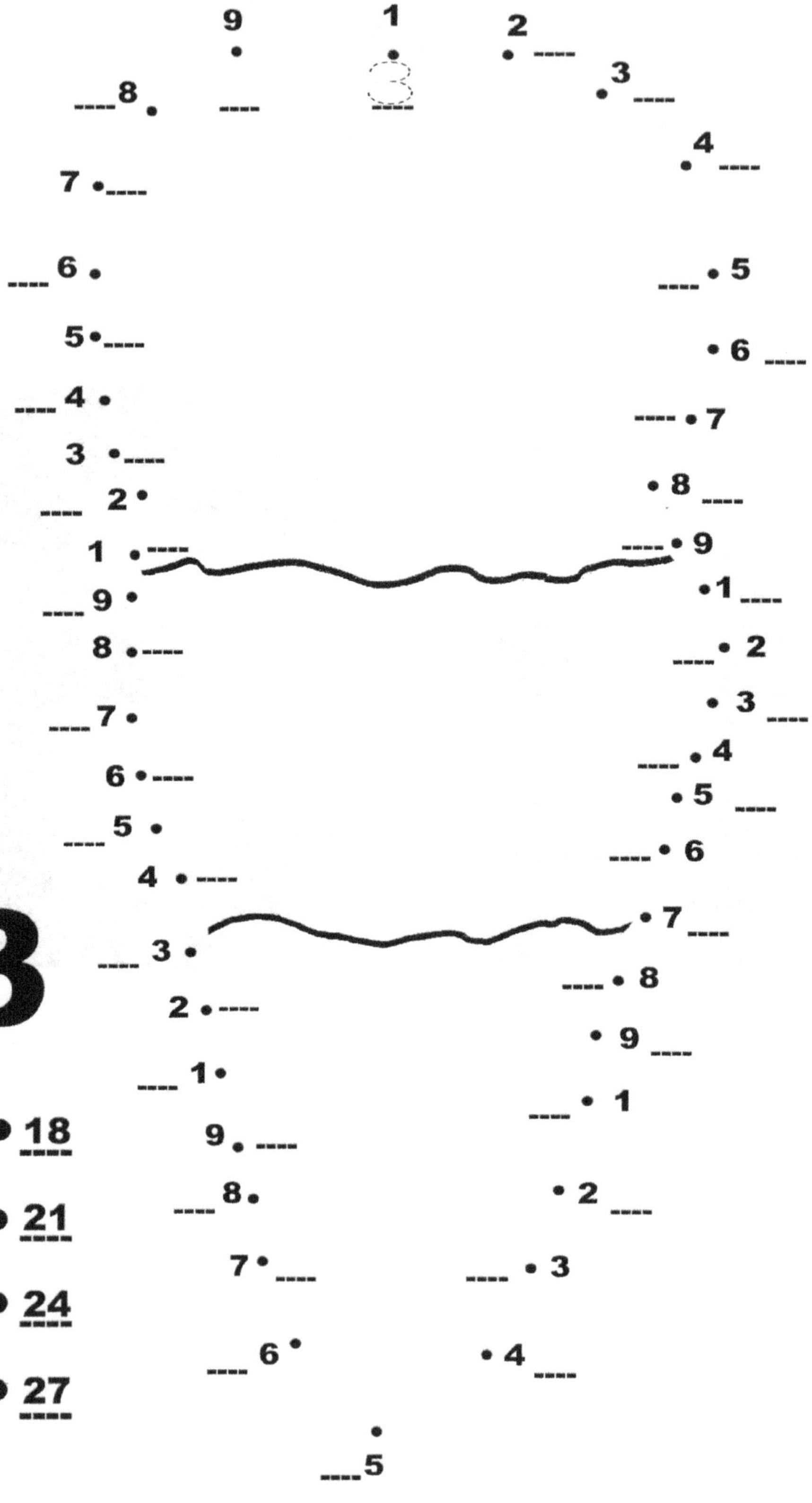

x3

1 • 3
2 • 6
3 • 9
4 • 12
5 • 15
6 • 18
7 • 21
8 • 24
9 • 27

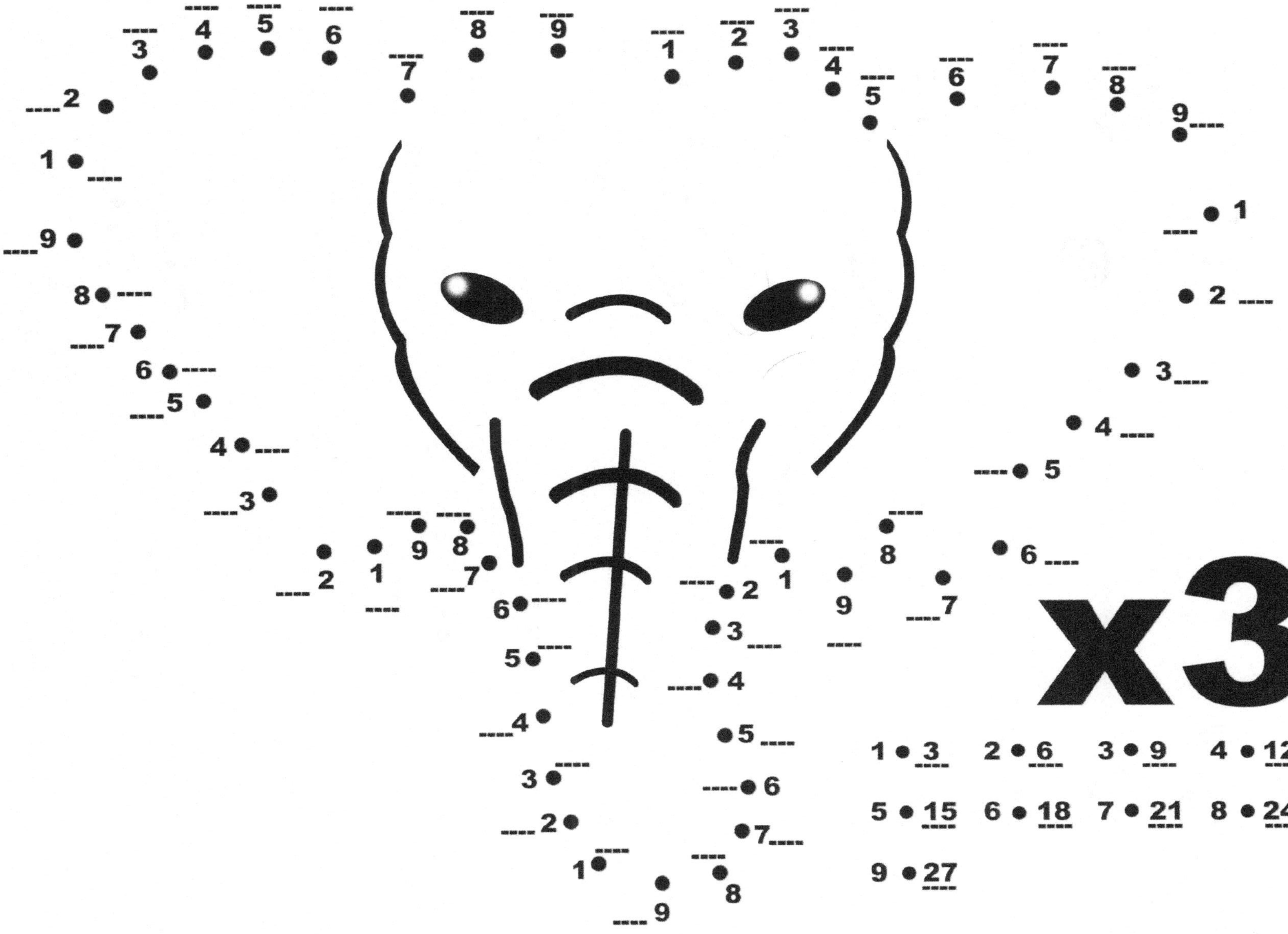

x3

1 • 3 2 • 6 3 • 9 4 • 12

5 • 15 6 • 18 7 • 21 8 • 24

9 • 27

x3

1 • 3

2 • 6

3 • 9

4 • 12

5 • 15

6 • 18

7 • 21

8 • 24

9 • 27

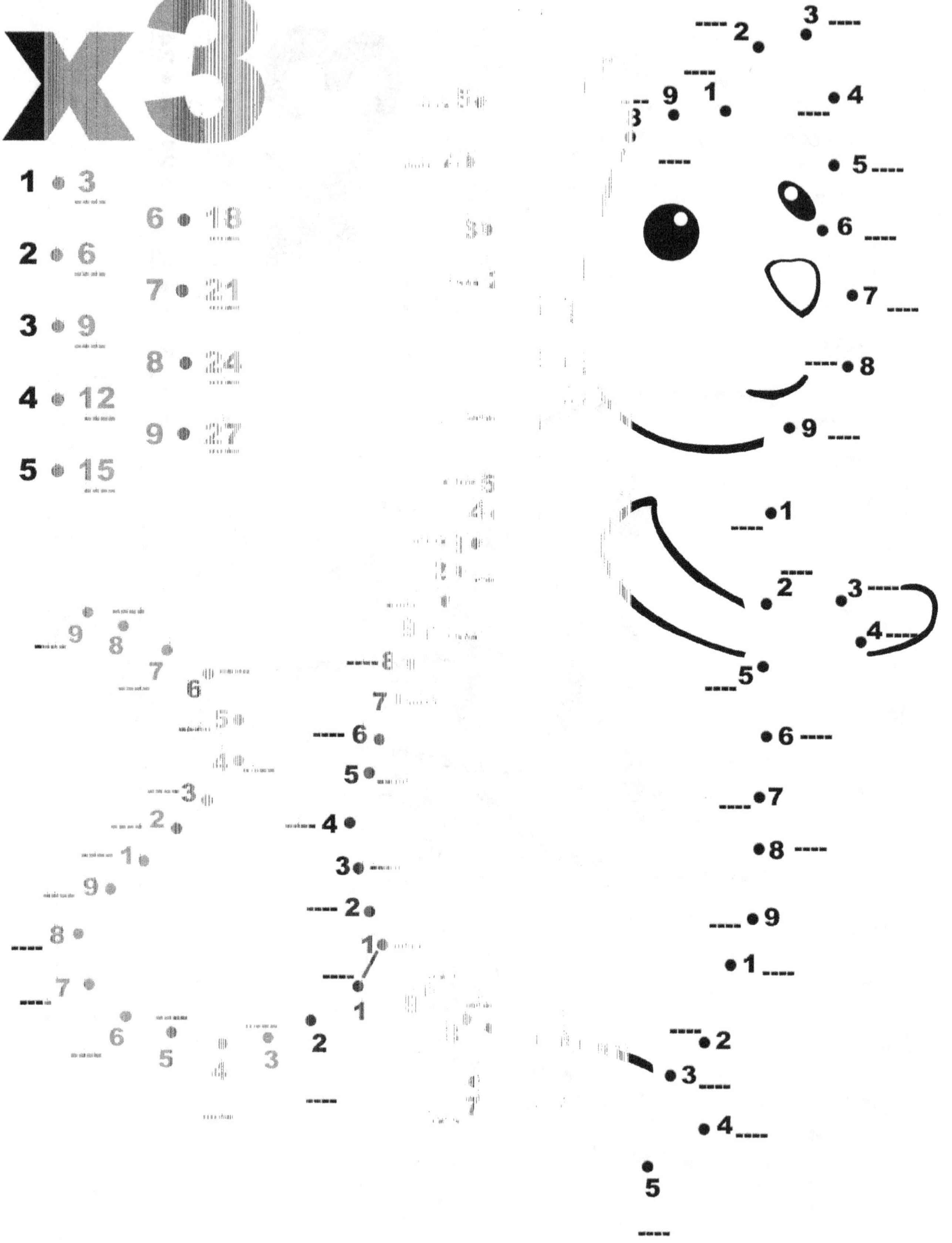

x3

1 • 3

2 • 6

3 • 9

4 • 12

5 • 15

6 • 18

7 • 21

8 • 24

9 • 27

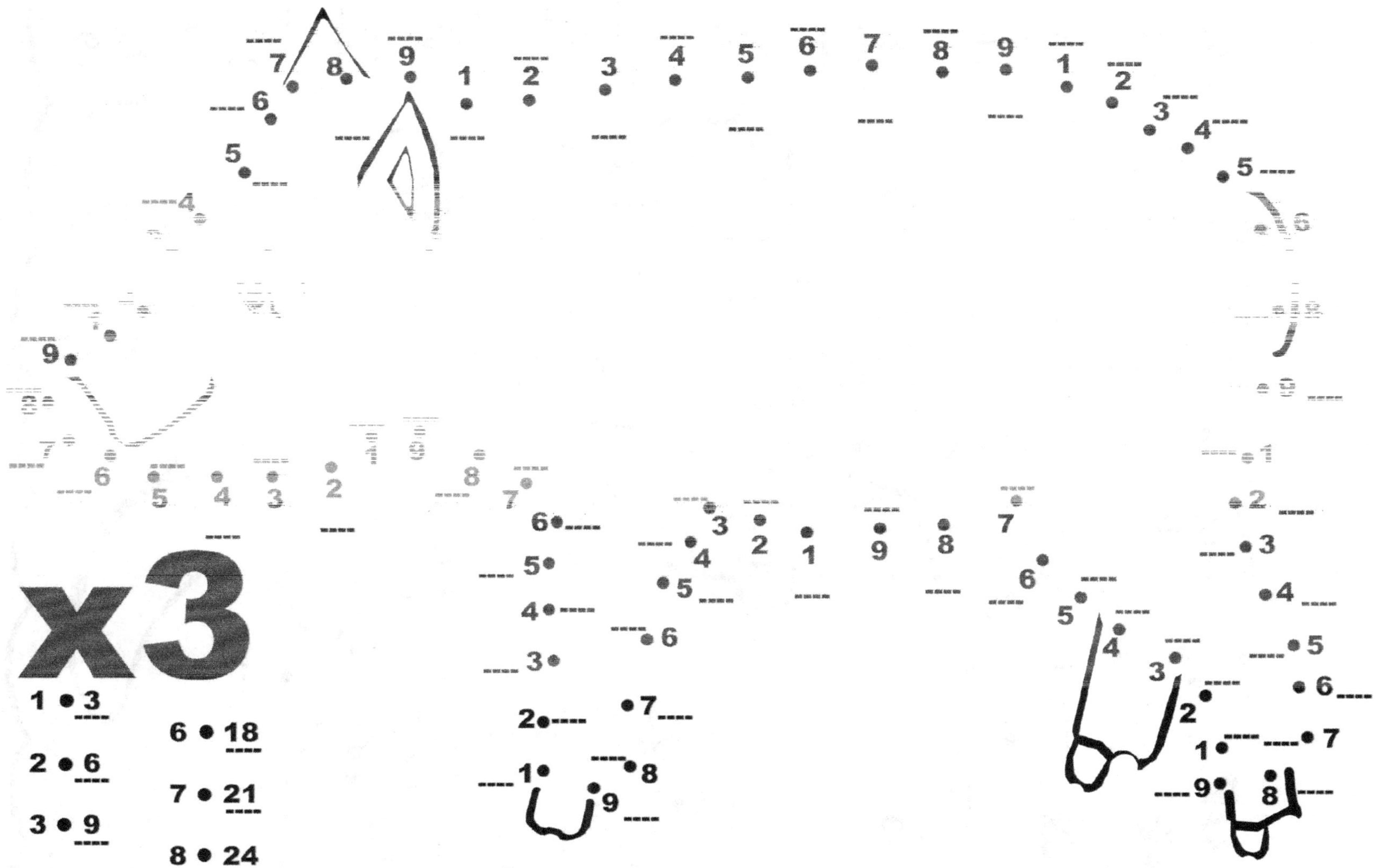

MIX UP LINE
Chili the Little Big Dog
x3
1 • 3
2 • 6
3 • 9
4 • 12
5 • 15
6 • 18
7 • 21
8 • 24
9 • 27

What did you learn?

3x1=3	3x1=	3x6=
3x2=6	3x2=	3x9=
3x3=9	3x3=	3x2=
3x4=12	3x4=	3x7=
3x5=15	3x5=	3x3=
3x6=18	3x6=	3x5=
3x7=21	3x7=	3x8=
3x8=24	3x8=	3x1=
3x9=27	3x9=	3x4=

Cover the answers and test yourself!

x 4

WARM UP

x4

1 • 4   2 • 8   3 • 12

4 • 16   5 • 20   6 • 24

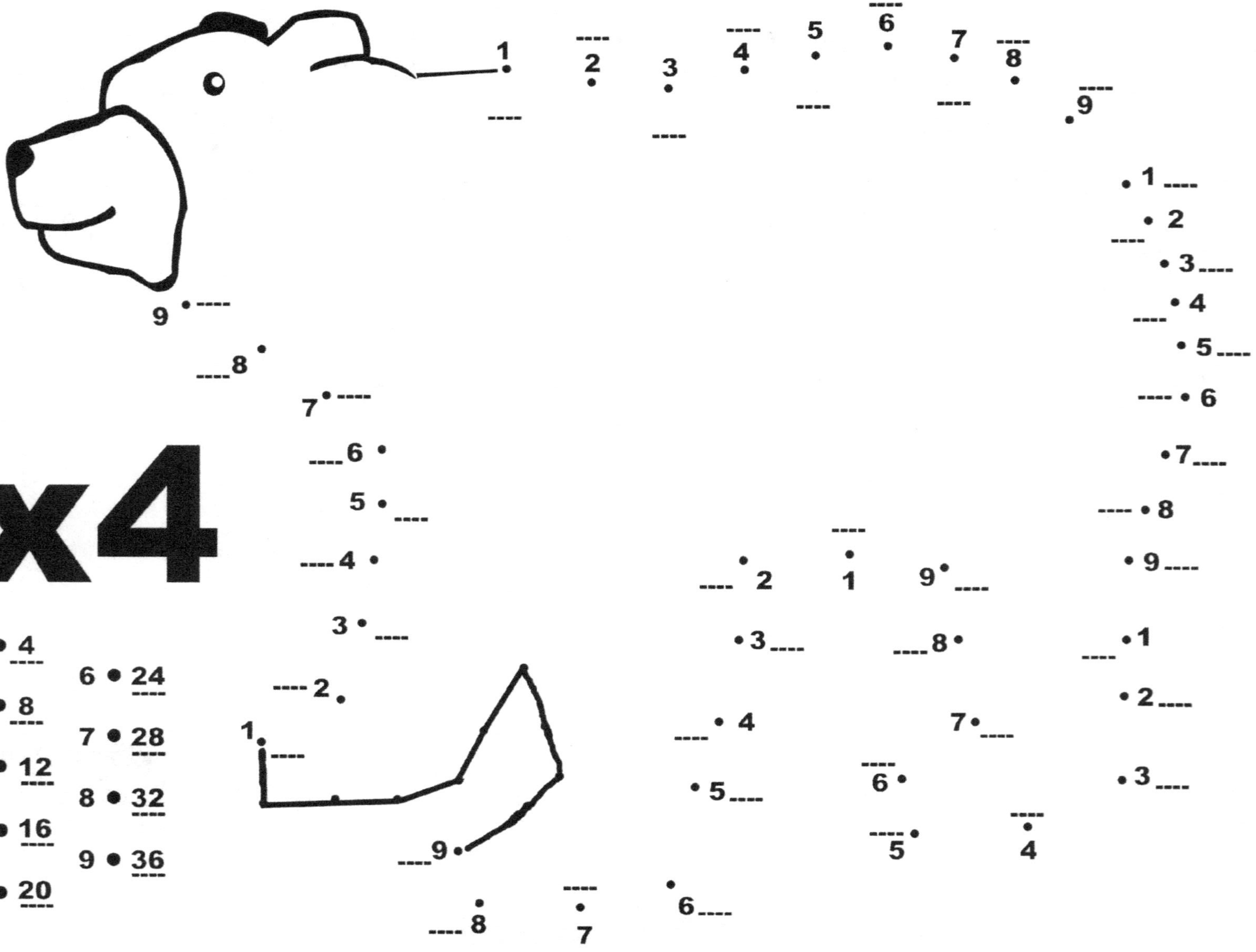

x4

1 • 4
2 • 8
3 • 12
4 • 16
5 • 20
6 • 24
7 • 28
8 • 32
9 • 36

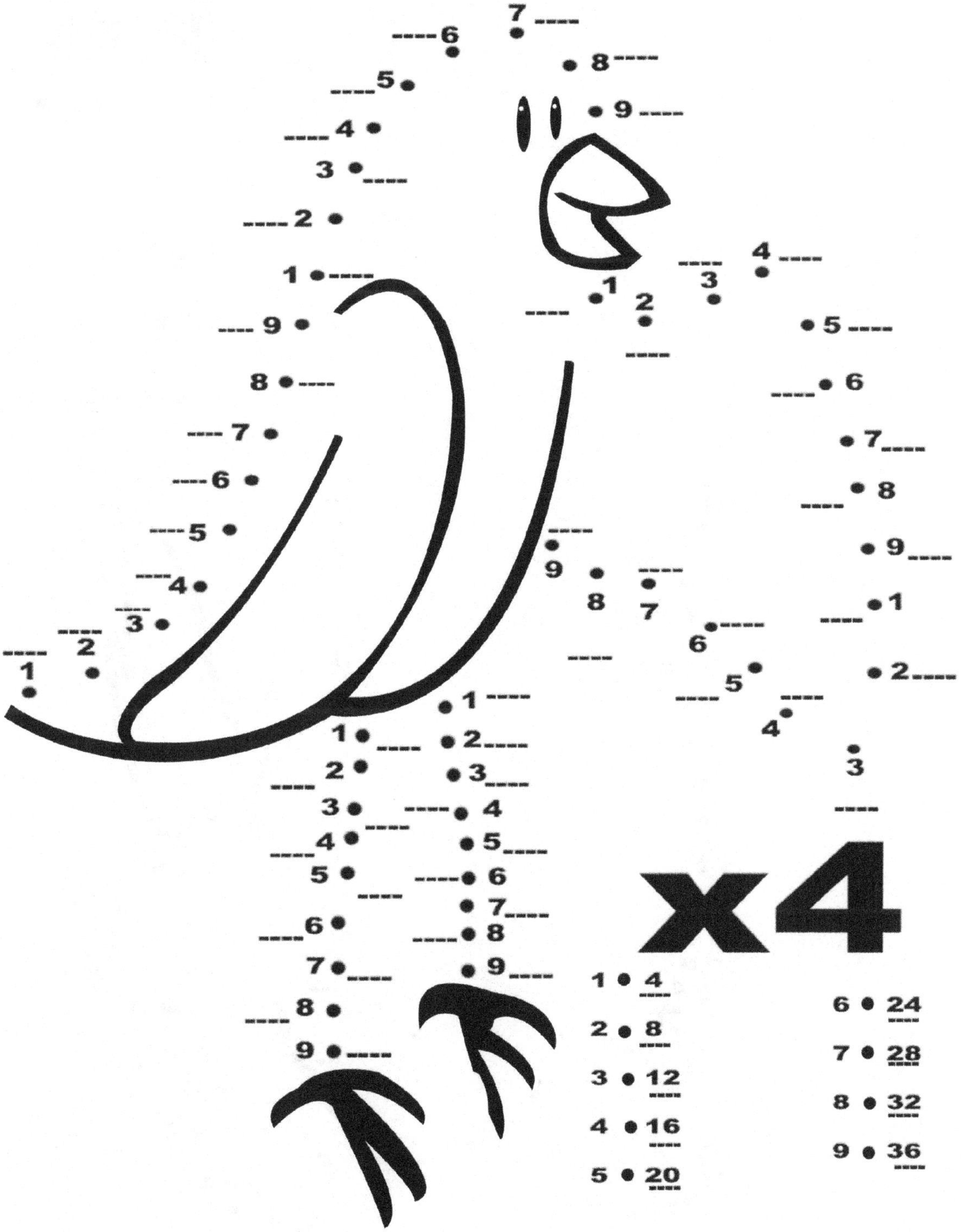
x4
1 • 4
2 • 8
3 • 12
4 • 16
5 • 20
6 • 24
7 • 28
8 • 32
9 • 36

x4

1 • 4   2 • 8

3 • 12   4 • 16

5 • 20   6 • 24

7 • 28   8 • 32

9 • 36

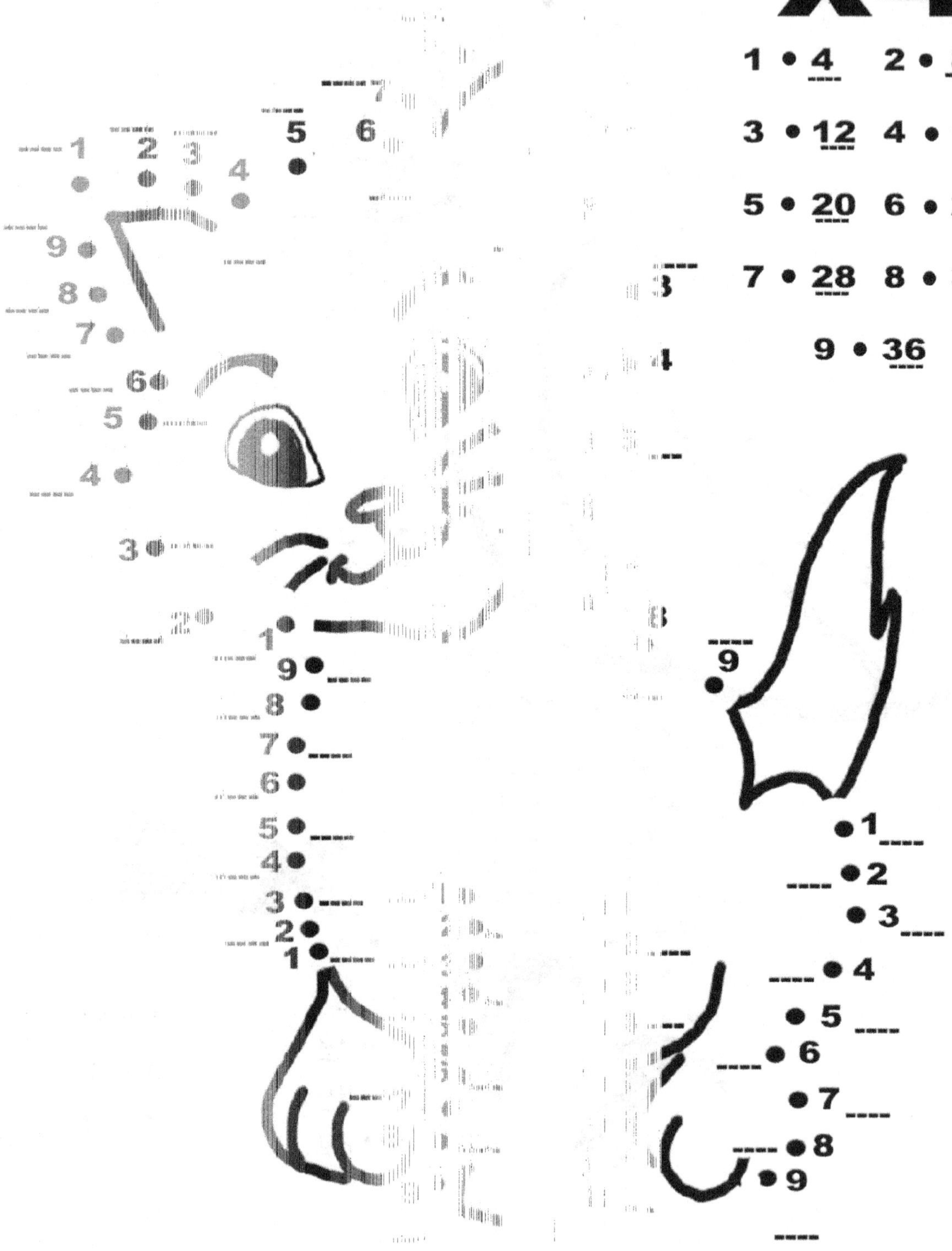

CHILI THE LITTLE BIG DOG

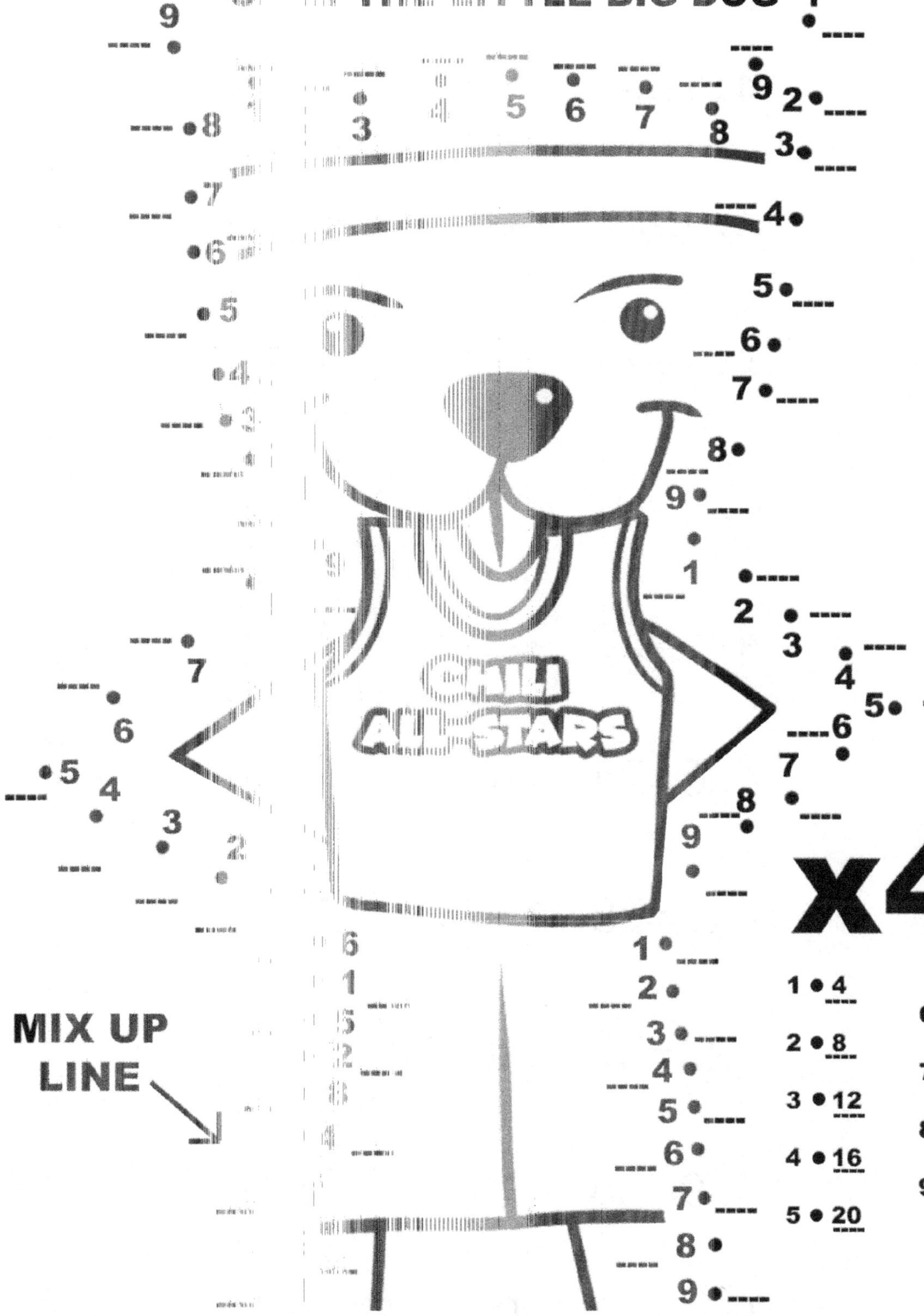

x4

1 • 4	6 • 24
2 • 8	7 • 28
3 • 12	8 • 32
4 • 16	9 • 36
5 • 20	

What did you learn?

4x1=4	4x1=	4x6=
4x2=8	4x2=	4x9=
4x3=12	4x3=	4x2=
4x4= 16	4x4=	4x7=
4x5=20	4x5=	4x3=
4x6=24	4x6=	4x5=
4x7=28	4x7=	4x8=
4x8=32	4x8=	4x1=
4x9=36	4x9=	4x4=

Cover the answers and test yourself!

x6

WARM

x6

1 • 6   2 •   4 • 24   5 • 30

6 • 36   8 • 48   9 • 54

x6

1 • 6	6 • 36
2 • 12	7 • 42
3 • 18	8 • 48
4 • 24	9 • 54
5 • 30	

x6

1 • 6
2 • 12
3 • 18
4 • 24
5 • 30
6 • 36
7 • 42
8 • 48
9 • 54

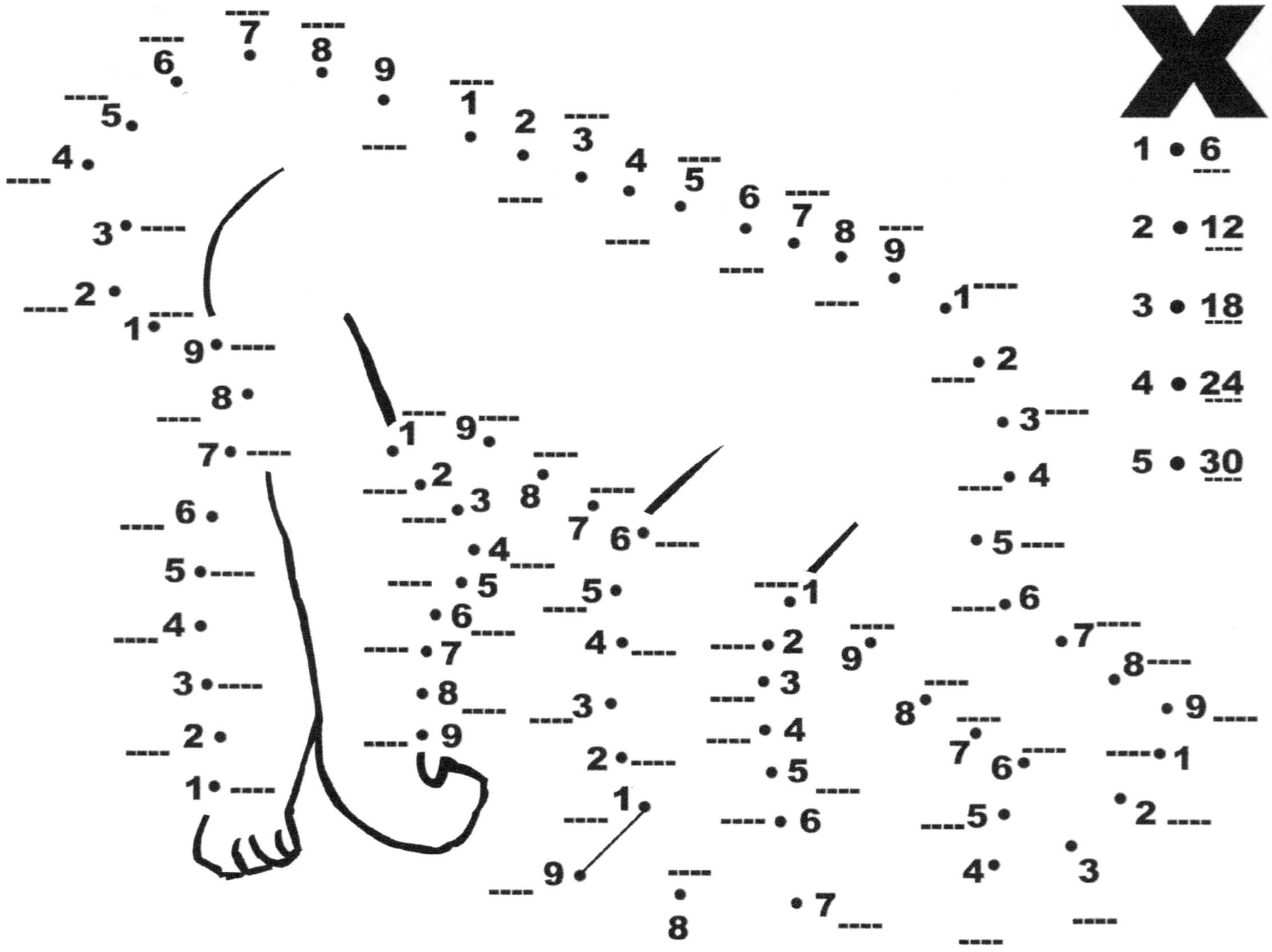

ROXY LA'ROCK

x6

1 • 6   2 • 12   3 • 18   4 • 24   5 • 30

6 • 36   7 • 42   8 • 48   9 • 54

CHILI TH… L… BIG DOG

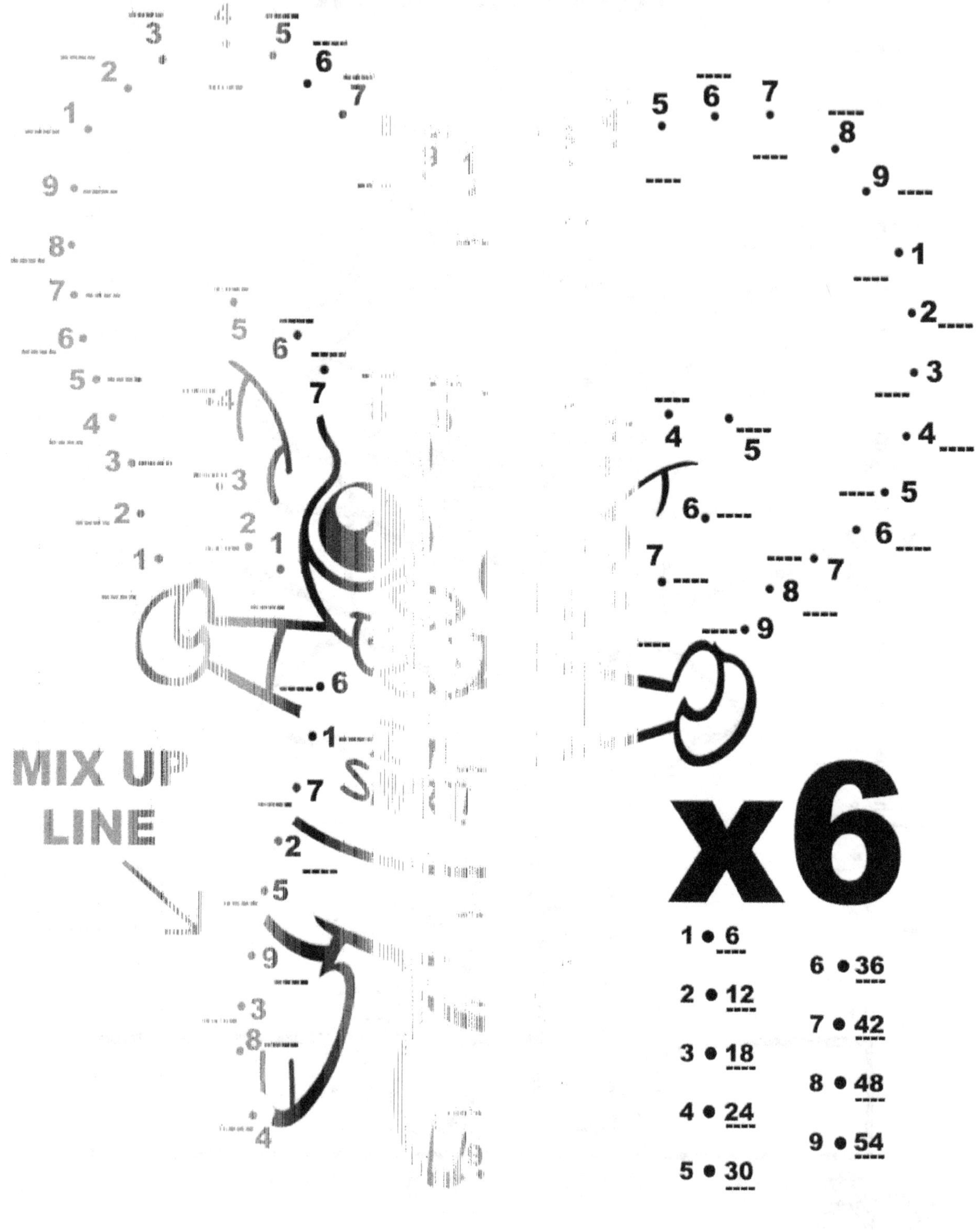

x6

1 • 6	6 • 36
2 • 12	7 • 42
3 • 18	8 • 48
4 • 24	9 • 54
5 • 30	

What did you learn?

6x1=6	6x1=	6x6=
6x2=12	6x2=	6x9=
6x3=18	6x3=	6x2=
6x4=24	6x4=	6x7=
6x5=30	6x5=	6x3=
6x6=36	6x6=	6x5=
6x7=42	6x7=	6x8=
6x8=48	6x8=	6x1=
6x9=54	6x9=	6x4=

Cover the answers and test yourself!

x7

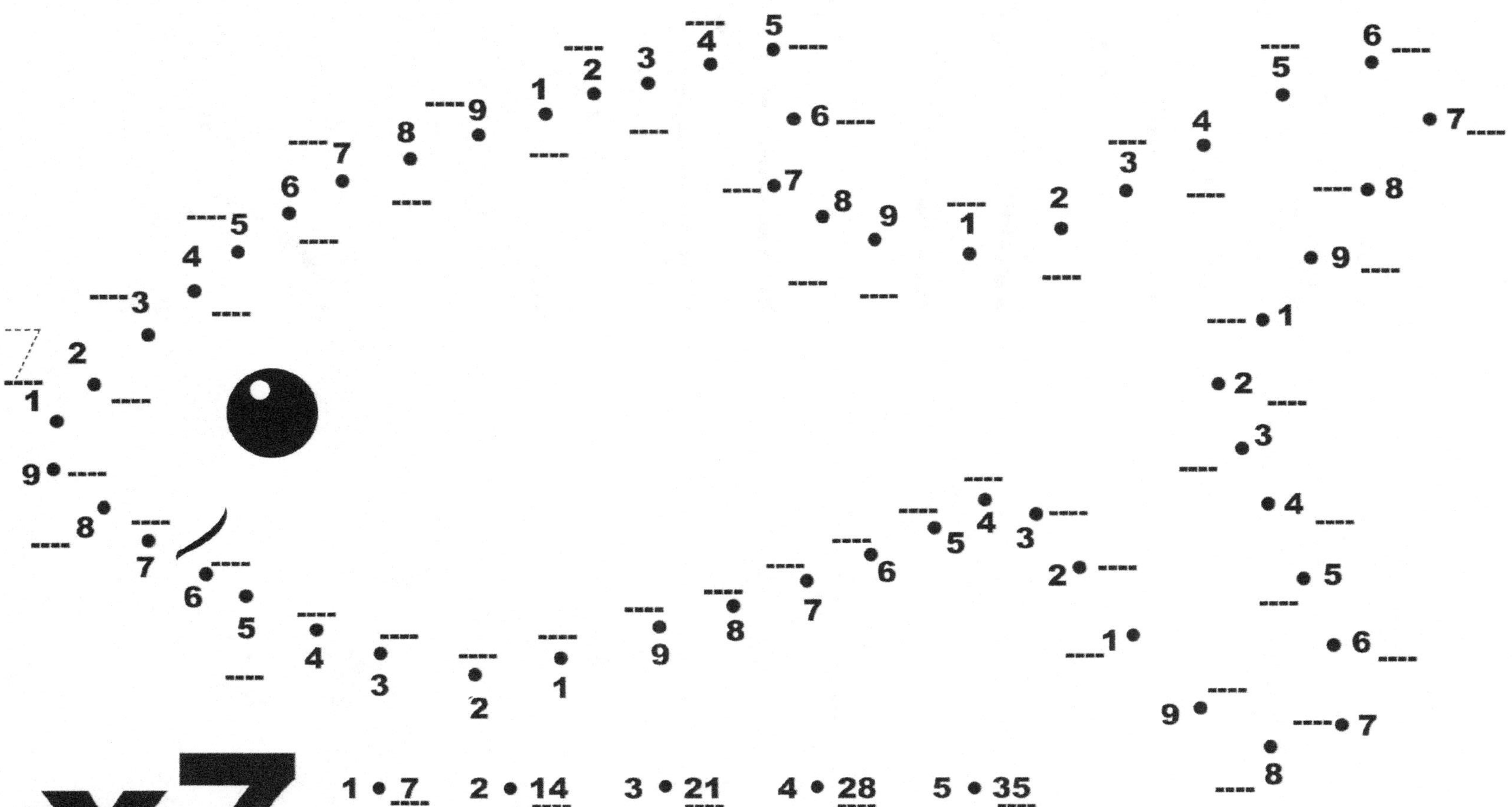

x7

1 • 7 2 • 14 3 • 21 4 • 28 5 • 35

6 • 42 7 • 49 8 • 56 9 • 63

x7

1 • 7
2 • 14
3 • 21
6 • 42
7 • 49
8 • 56

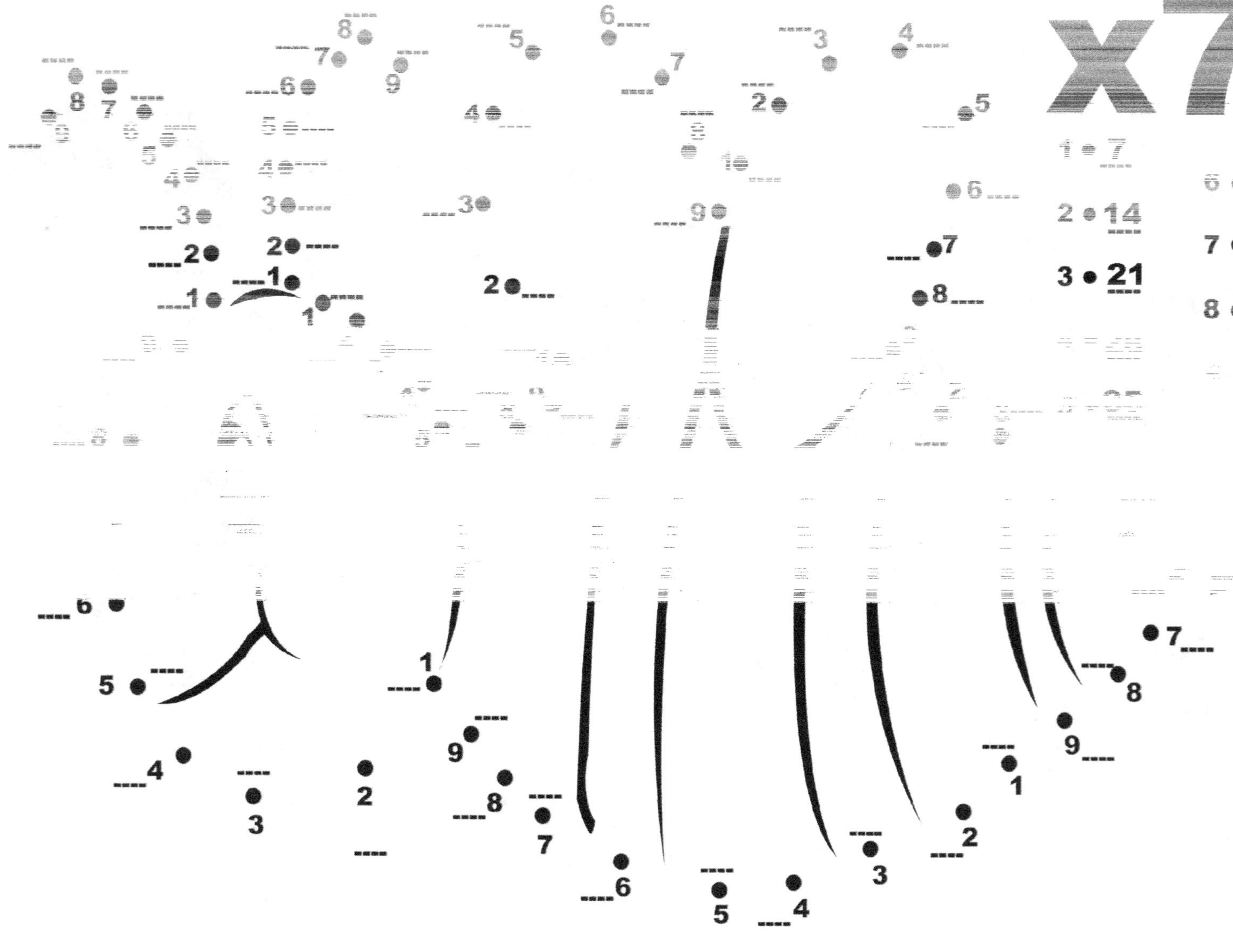

x7

1 • 7 ____

2 • 14 ____

3 • 21 ____

4 • 28 ____

5 • 35 ____

6 • 42 ____

7 • 49 ____

8 • 56 ____

9 • 63 ____

x7

1 • 7
2 • 14
3 • 21
4 • 28
5 • 35
6 • 42
7 • 49
8 • 56
9 • 63

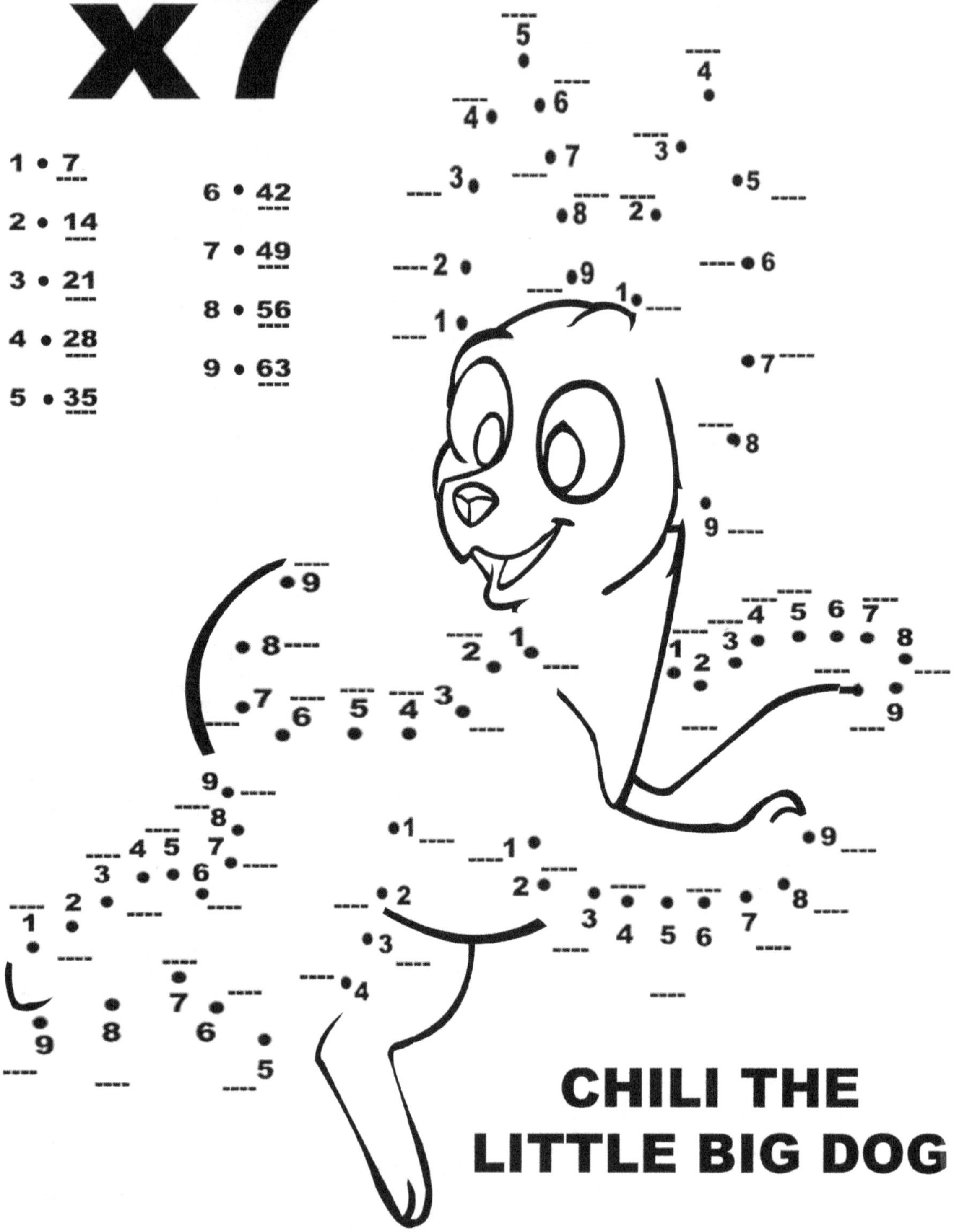

CHILI THE LITTLE BIG DOG

x7

1 • 7	6 • 42
2 • 14	7 • 49
3 • 21	8 • 56
4 • 28	9 • 63
5 • 35	

MIX UP LINE

CALI CO'CAT

What did you learn?

7x1=7
7x2=14
7x3=21
7x4=28
7x5=35
7x6=42
7x7=49
7x8=56
7x9=63

7x6=
7x9=
7x2=
7x7=
7x3=
7x5=
7x8=
7x1=
7x4=

Cover the answers and test yourself!

X8

WARM UP

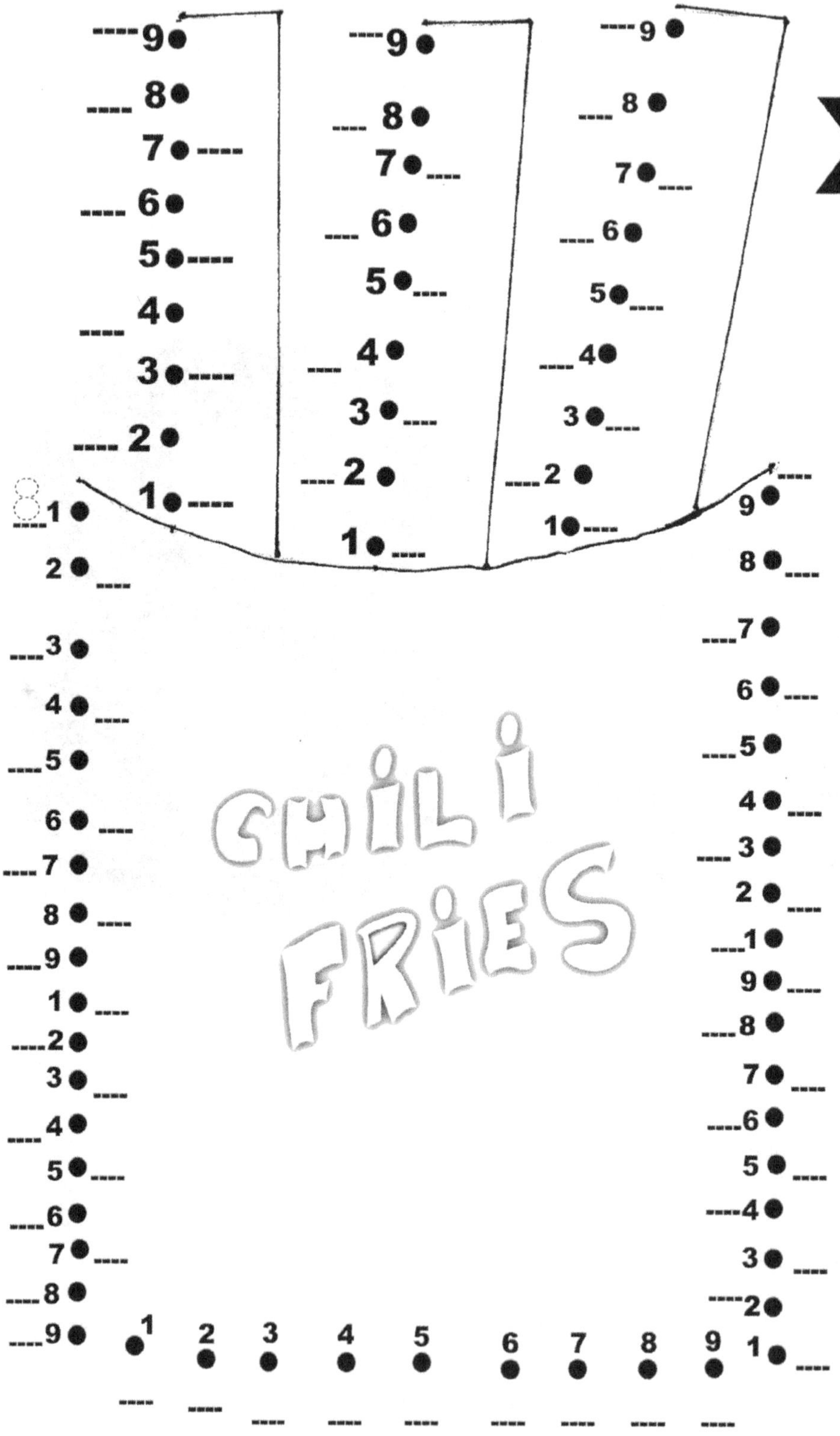

x8

1 • 8

2 • 16

3 • 24

4 • 32

5 • 40

6 • 48

7 • 56

8 • 64

9 • 72

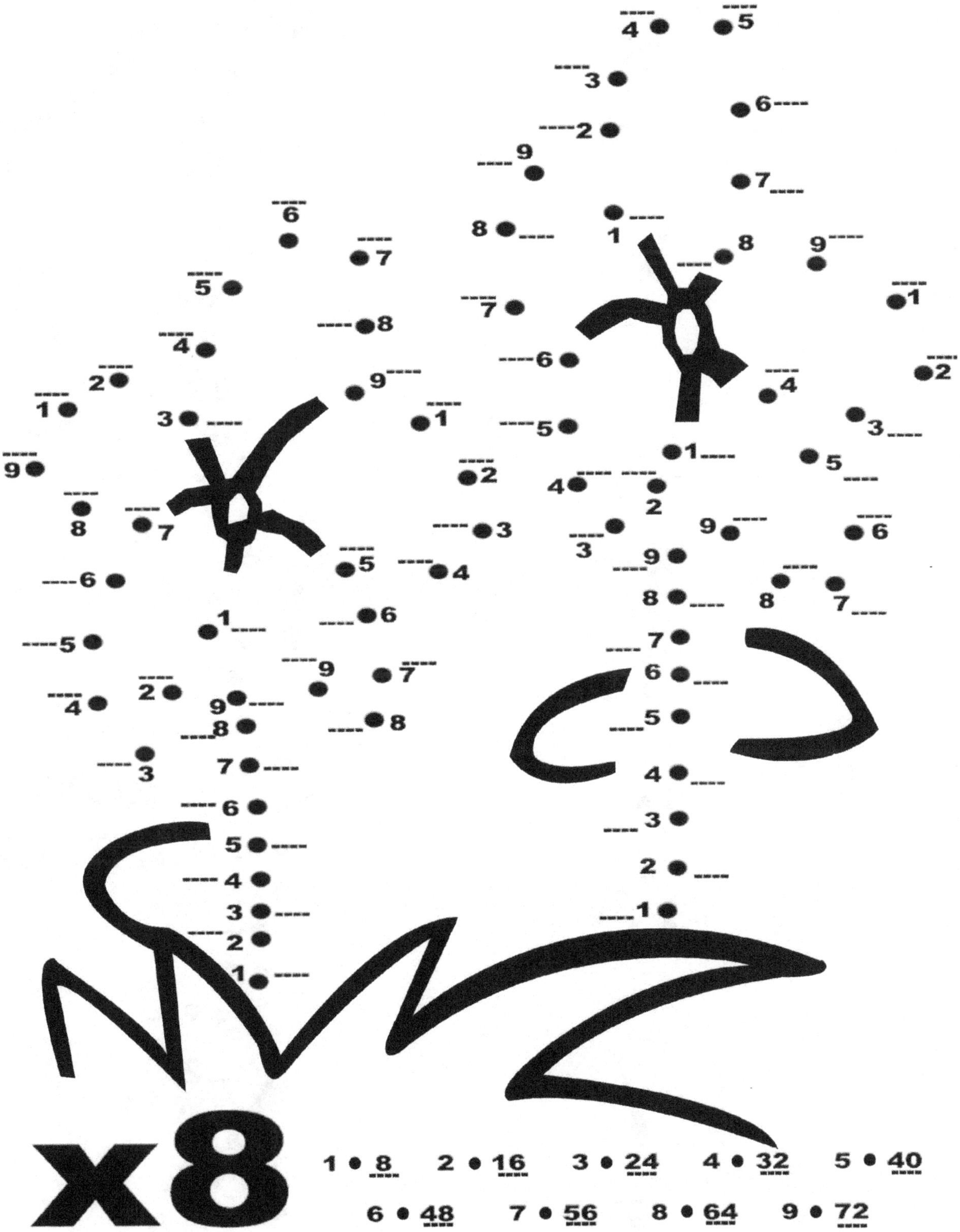

x8

1 • 8 2 • 16 3 • 24 4 • 32 5 • 40

6 • 48 7 • 56 8 • 64 9 • 72

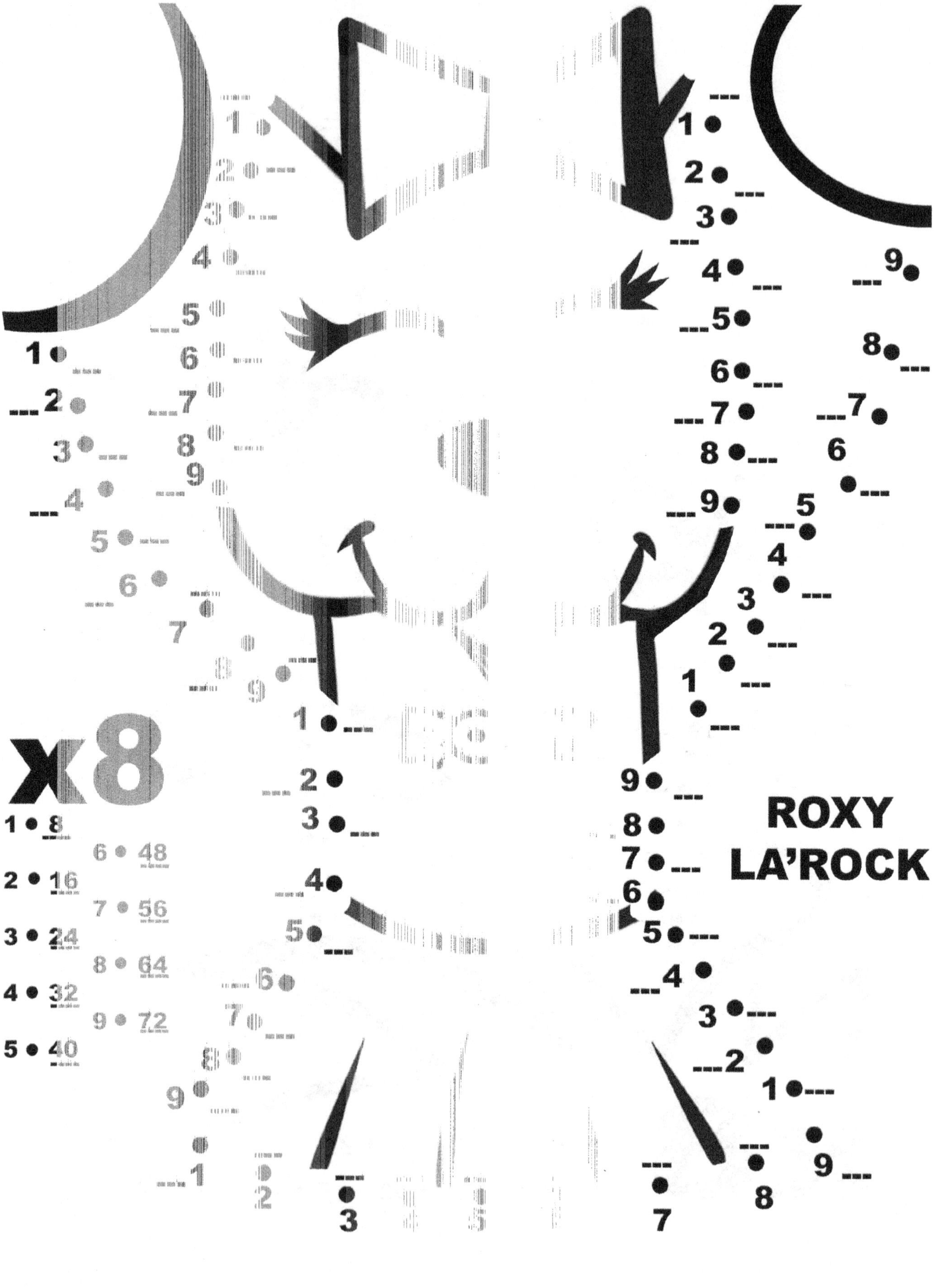

x8

1 • 8
2 • 16
3 • 24
4 • 32
5 • 40
6 • 48
7 • 56
8 • 64
9 • 72

ROXY
LA'ROCK

CHILI
THE LITTL
BIG DOG

x8

1 • 8

2 • 16

3 • 24

4 • 32

5 • 40

6 • 48

7 • 56

8 • 64

9 • 72

x8

1 • 8

2 • 16

3 • 24

4 • 32

5 • 40

6 • 48

7 • 56

8 • 64

9 • 72

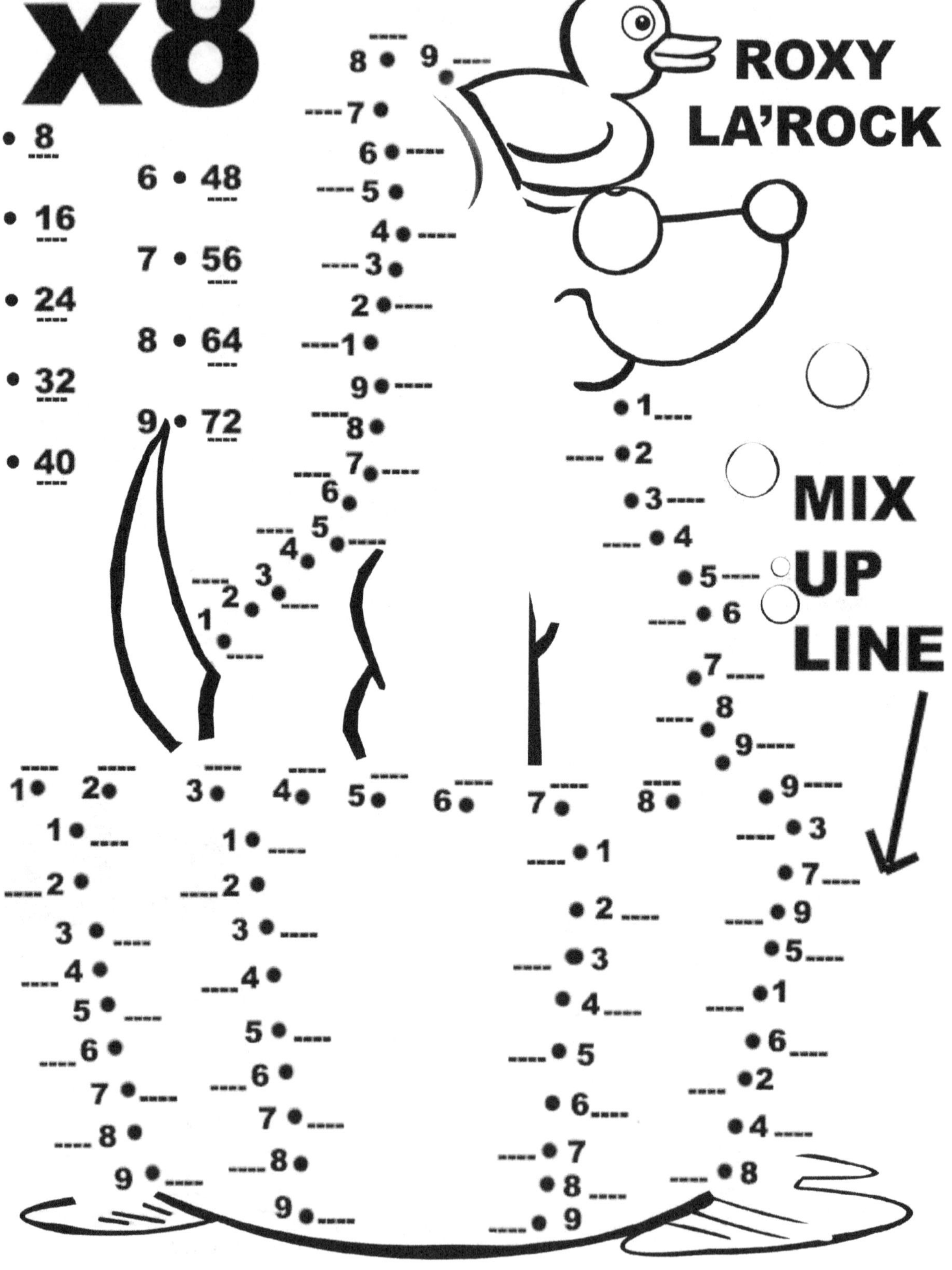

What did you learn?

8x1=8	8x1=	8x6=
8x2=16	8x2=	8x9=
8x3=24	8x3=	8x2=
8x4=32	8x4=	8x7=
8x5=40	8x5=	8x3=
8x6=48	8x6=	8x5=
8x7=56	8x7=	8x8=
8x8=64	8x8=	8x1=
8x9=72	8x9=	8x4=

Cover the answers and test yourself!

x 9

WARM UP

x9

1 • 9   2 • 18   3 • 27   4 • 36   5 • 45

6 • 54   7 • 63   8 • 72   9 • 81

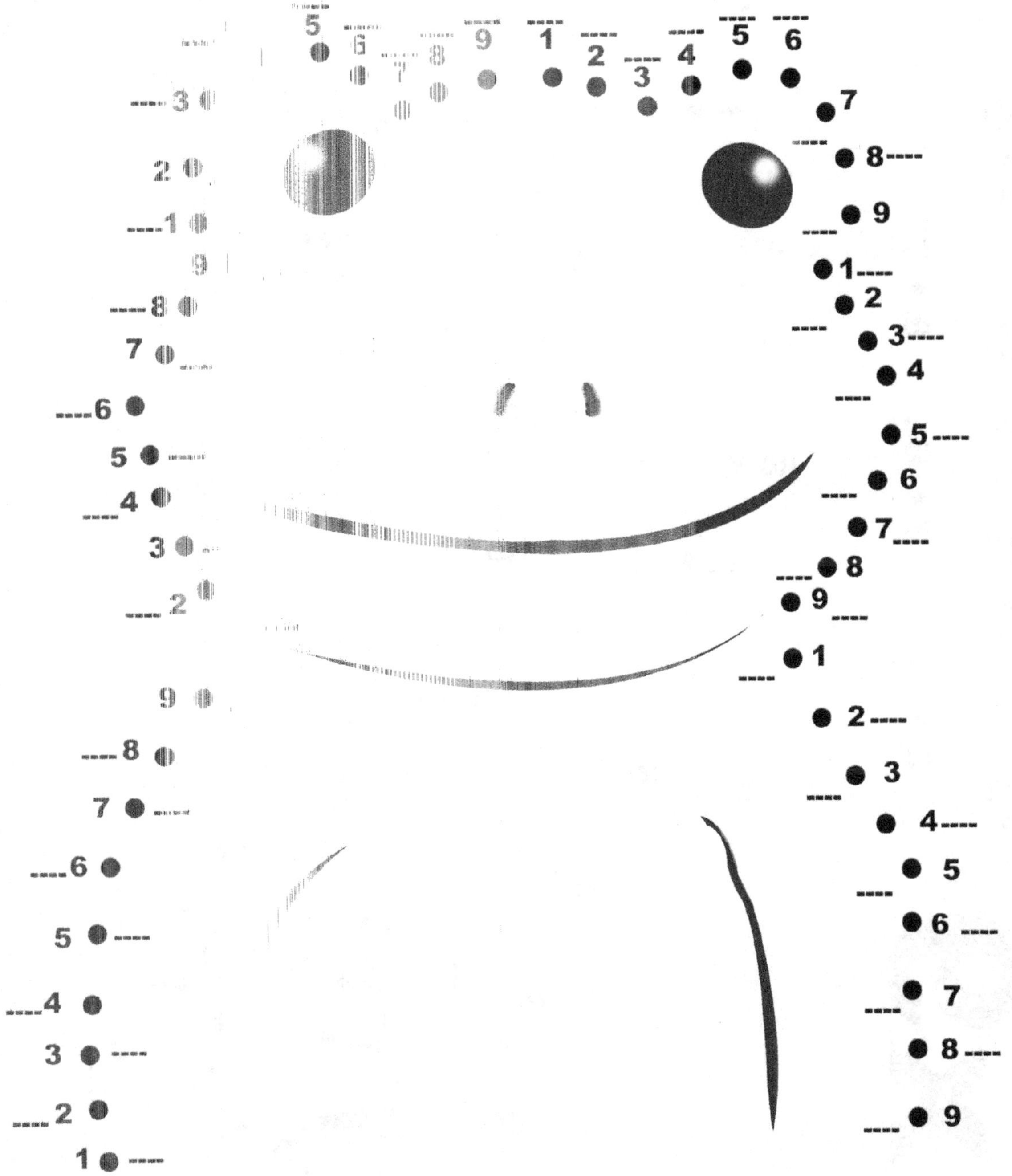

x9

1 • 9 | 2 • 18 | 3 • 27 | 4 • 36 | 5 • 45

6 • 54 | 7 • 63 | 8 • 72 | 9 • 81

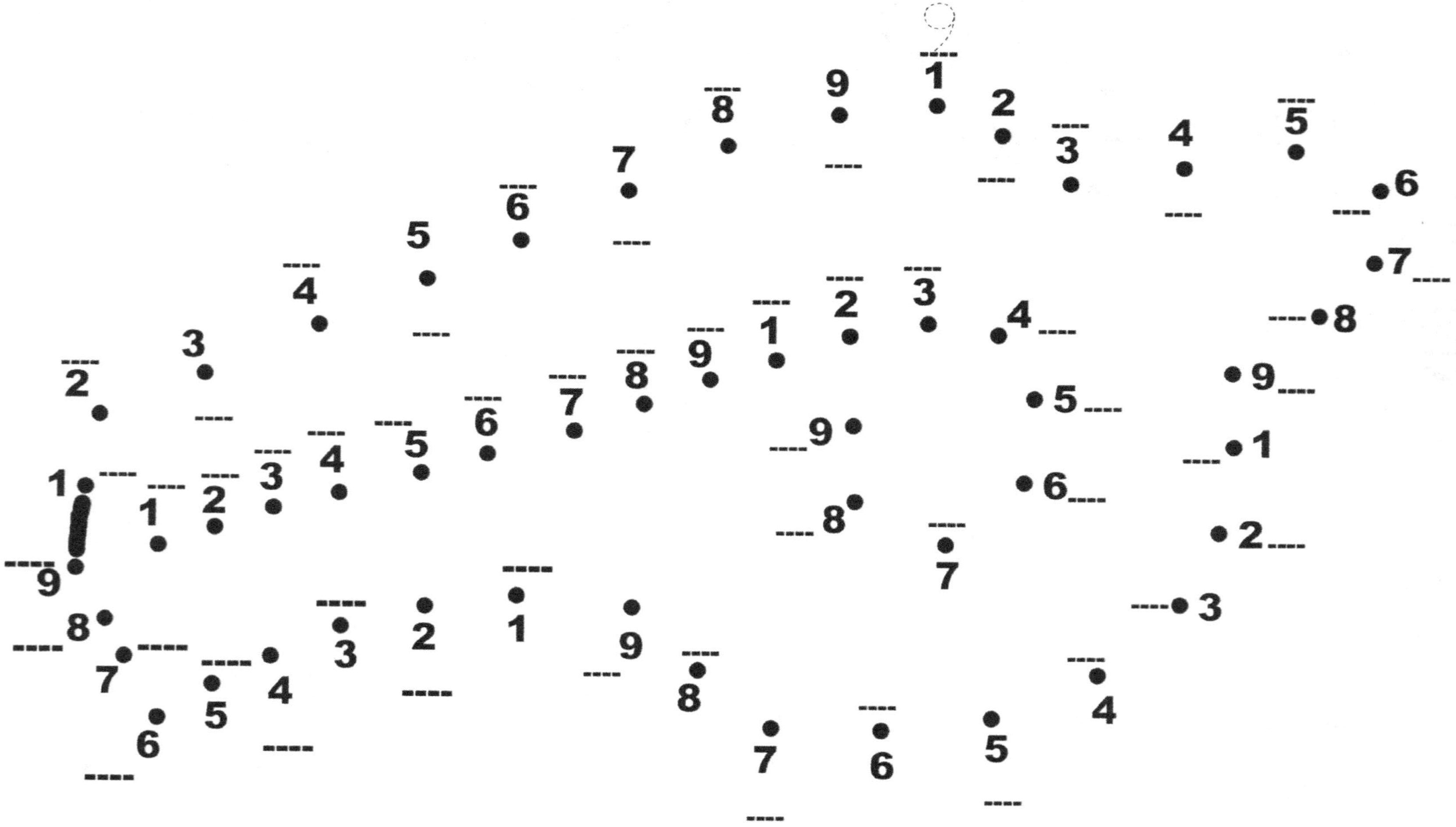

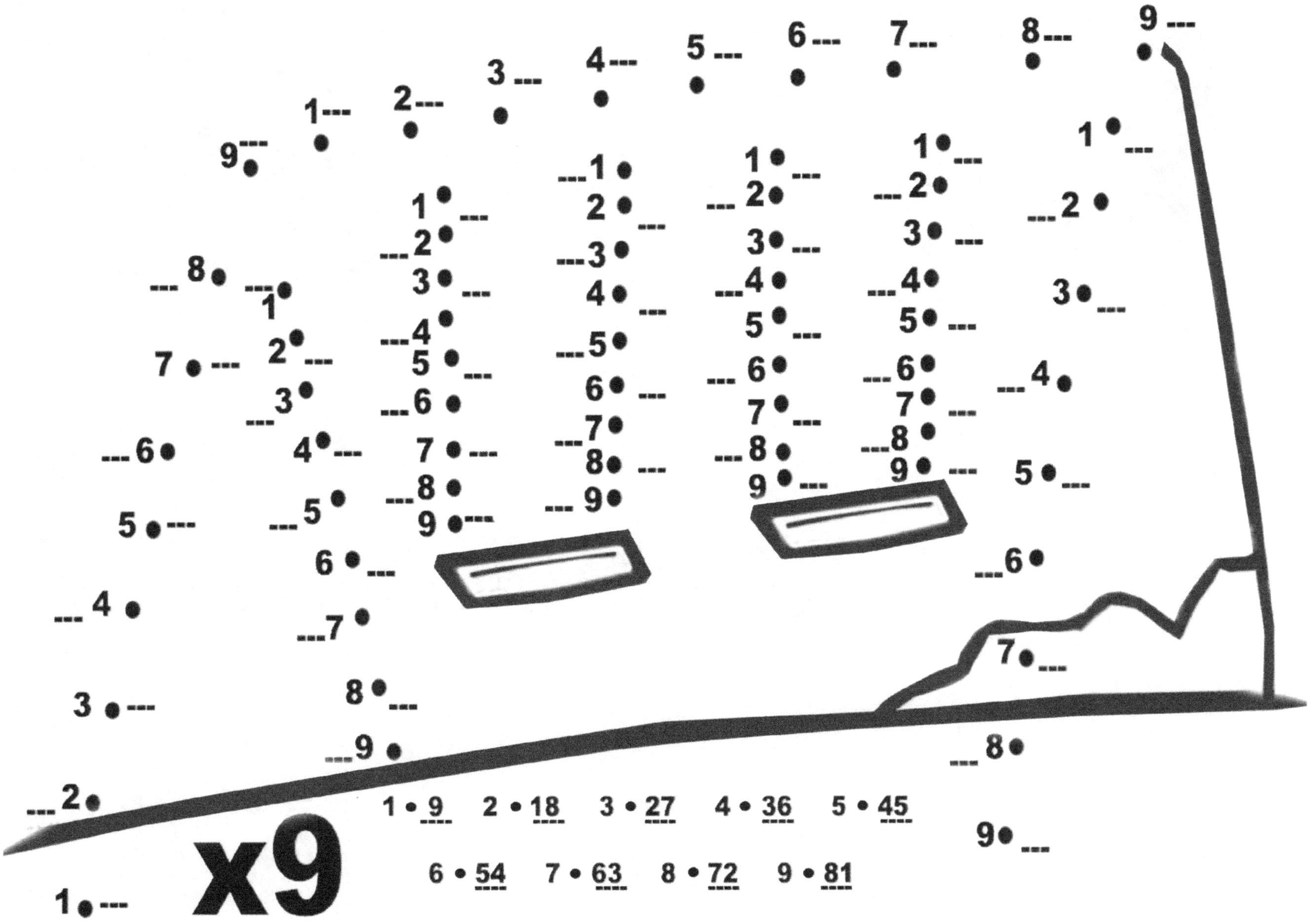

x9

1 • 9　2 • 18　3 • 27　4 • 36　5 • 45

6 • 54　7 • 63　8 • 72　9 • 81

x9

1 • 9
2 • 18
3 • 27
4 • 36
5 • 45
6 • 54
7 • 63
8 • 72
9 • 81

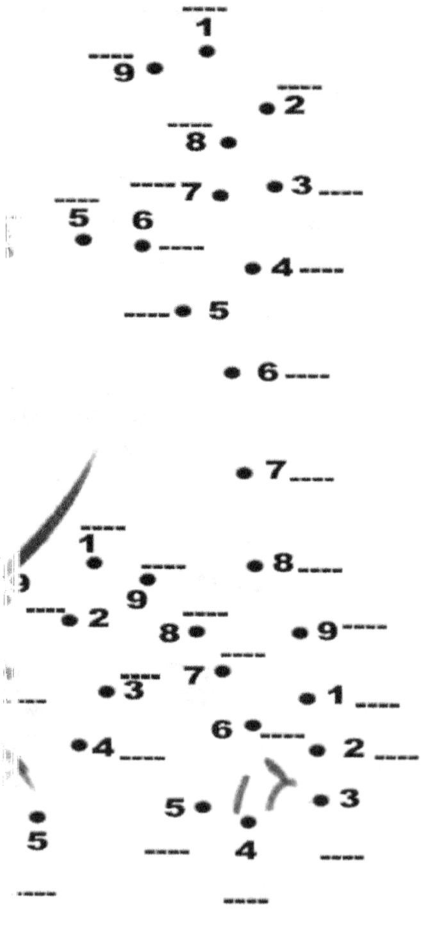

CHILI
THE
LITTLE
BIG
DOG

ROXY LA'ROCK

x9

1 • 9
2 • 18
3 • 27
4 • 36
5 • 45
6 • 54
7 • 63
8 • 72
9 • 81

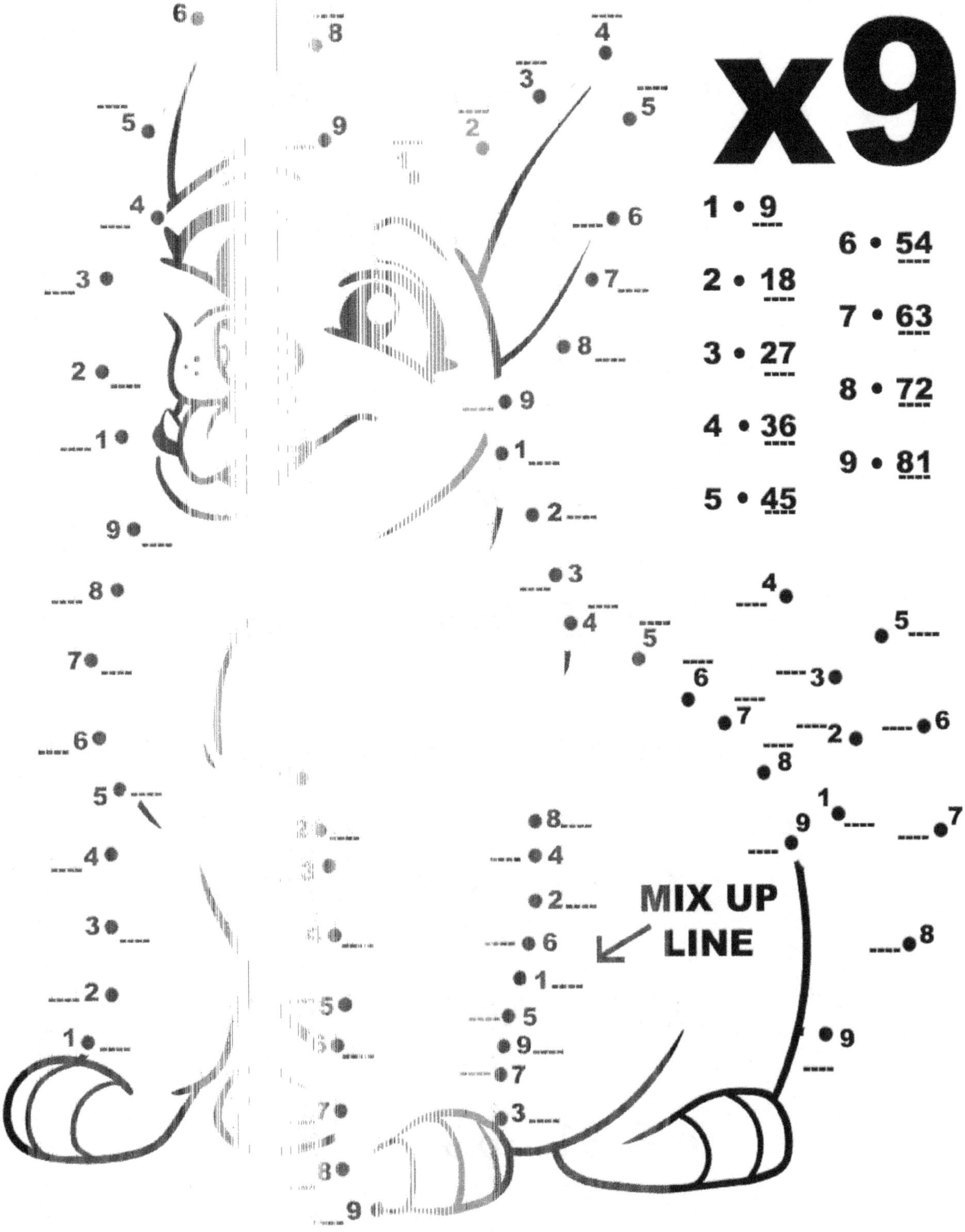

What did you learn?

9x1=9	9x1=	9x6=
9x2=18	9x2=	9x9=
9x3=27	9x3=	9x2=
9x4= 36	9x4=	9x7=
9x5=45	9x5=	9x3=
9x6=54	9x6=	9x5=
9x7=63	9x7=	9x8=
9x8=72	9x8=	9x1=
9x9=81	9x9=	9x4=

Cover the answers and test yourself!

Made in the USA
Monee, IL
05 November 2026

67382737R00031